371·1024

Lacock
Chippenham
Wiltshire

Motivating the Disaffected

Staying Safe

Dr Gerald Lombard

Staying Safe

Published by Lifetime Careers Wiltshire, 7 Ascot Court, White Horse Business Park, Trowbridge BA14 0XA.

© Gerald Lombard 2004

ISBN 1 902876 72 5

No part of this publication may be copied or reproduced, stored in a retrieval system or transmitted in any form or by any means electronic or mechanical or by photocopying or recording without prior permission of the publishers.

Printed by Cromwell Press, Trowbridge
Cover illustration by Russell Cobb
Text design by Ministry of Design

Motivating the Disaffected
Series editor: Dr Gerald Lombard

Staying Safe is one of a series of six titles designed to help professionals in education and advisory work to motivate and encourage students who are disengaged from learning.

Each book provides a concise and practical guide to topics that are of particular concern to teachers and advisers.

The other titles in the series are:

The ABC of Approach to Classroom Behaviour Management
Asperger Syndrome and high fuctioning autism:
    guidelines for education post 16
Complex Specific Learning Difficulties
Motivational Triggers
Social Competence: reading other people

To order copies, please contact Orca Book Services Ltd, Stanley House, 3 Fleets Lane, Poole, Dorset BH15 3AJ. Tel: 01202 665432. Fax: 01202 666219.

For further information about these and other products published by Lifetime Careers Publishing, please contact our customer services, tel: 01225 716023; email: sales@lifetime-publishing.co.uk, or
www.lifetime-publishing.co.uk

Dr Gerald Lombard, C. Psychol., AFBPsS

Ged is Director of The Independent Psychological Service, which is an intervention and training service for young people and adults. He primarily works with individuals who 'won't, can't or can't be arsed' (one client's view of their work).

As a Chartered Psychologist, his major areas of interest are motivational principles, social competency (reading faces with intent) and complex specific learning difficulties. Ged was a secondary school teacher for 15 years, a part-time tutor/psychologist at two prisons and has held his current post for over ten years.

Contents

Introduction

I am an individual who firmly believes in the Maslow approach to motivation and survival:

- physiological needs come first (food, water)
- safety needs come next
- then, I can function socially.

The following outlines my basic principals of staying safe on a daily basis: at home, at work, travelling between destinations. It begins with some of the worst situations you are likely to meet, and concludes with how to avoid entering unsafe zones and circumstances at work.

I am grateful to all professionals who have enabled me over the years to work confidently with a wide range of individuals – and to my family who have taught me the most.

I cannot take any responsibility for another person's safety, based on the words that follow. I can only offer my views on my personal approach to my attempts at remaining unharmed, and reducing the risks to those I work with.

Maybe I have been lucky so far. At work, with individuals and groups, I have usually felt relatively safe and secure. Admittedly, I cannot always be as confident about my effectiveness or value!

My work brings me into contact with a variety of young adults in different settings: in residential or secure units, colleges, clinics, or their own homes. From my earliest professional experiences, I have made safety (mine and others) a primary issue. Unless I, and those I work with, feel safe the subsequent aspects of any relationship or situation are unlikely to develop positively.

Therefore, what follows are the key aspects of what I believe have kept me safe, professionally and personally... so far! It is not a guaranteed guide for any or all situations, but specific aspects may help the planning of some activities or help in certain organisations. I like to believe that I am still willing to take risks with the clients I work with – but in a responsible domain where the percentages are likely to be high in the staying safe column.

Section 1
Conflict resolution: the threatening aspects

Avoiding danger

You should always trust your feelings. Even if you cannot prove it, or see a definite sign, if you sense something is wrong then something is likely to be wrong. We have all evolved over millions of years. We are primed and programmed to sense danger. Our more comfortable lifestyles and more consistent environments have made us blasé to some dangers – but our instincts can still pick up signals we cannot consciously understand. If these signals do occur, then try to take some action that will increase your safety. For example:

- get someone else to join you
- if alone with one other person, hurry the consultation, or in extreme circumstances:
- abandon the situation.

Remember, 'a signal of danger' may only be a feeling. For me, it can be a change in atmosphere, the hairs raising on the back of my neck – with no explanation – but I take notice of it.

The worst case... outside work

If physically challenged, you should already know what your personal strengths are for such a crisis. I am no longer athletic or nimble – more pathetic and a namby-pamby. However, I can talk – ask anyone who knows me! So, I know that in a threatening situation I would use my calming, but assertive, voice. I can no longer sprint because of my arthritis, nor can I confidently fight and kick. I have done both in my youth when confronted by others' physical aggression. Nowadays, I know that my crucial skill – facing a potential violent attack – is to use my negotiation skills (they must be up to scratch – I have used them successfully several times with police in armed-siege incidents).

Have you already audited your key skills if physically challenged? Would you, like me, have to rely initially on 'talk down' to the potential aggressor? I believe most middle-aged people would. However, I do not discount that many people's reactions would be to (a) sprint (because they are fast runners), or (b) physically attack at the threat of violence (because they are physically strong enough to do that). The question is this – which method(s) are you most capable of? Decide now. Making your mind up now will help when you are in a situation where you will need to know.

Unfortunately, the threat of violence occasionally spirals to physical attack. Probably the best advice is that from self-defence experts and that of Piven and Borgenicht (2001). If wanting to survive, say, a mugging:

- Do not argue or fight with a mugger unless your life is in danger. If they try to grab your wallet, purse, case, portable computer, handbag – let them have them – your life is worth more.
- If it is very clear they want to harm you, and an attack is imminent or is happening, experts (I'm not one!) advise:
  - aim to disable them with your first thrust/blow by attacking sensitive/vital areas of the attacker's body, e.g. thrust fingers into and above the attacker's eyes
  - drive knee in upward direction into groin, grab and crush testicles (if male attacker)

- hit the front of the attacker's throat, with the side of your hand, held straight and strong
- angle the tip of your elbow, with real force, into the side of the attacker's ribs
- stamp down on the attacker's instep.

- Use available objects as weapons:
  - keys can be used between fingers to jab, punch, slash
  - pick up any available object e.g. board/card, piece of wood, detachable items from vehicles (aerial/wipers) – any available item could be used as a deterrent or potential weapon.

It all sounds incredibly violent – but this is the last resort. For your own safety, decide what aspects described above would best serve you. If you wait until the incident happens, you are likely to 'freeze', i.e. not able to think or function, and become more of a victim.

States, or levels of awareness

These are essential personal ingredients in managing yourself in a variety of personal, social and occupational situations.

For example, while I am writing these words, I am in a comfortable situation, or 'comfort zone'. I am not aware that I require anything more than to merely sit, concentrate on the words and the page, and occasionally move to get a coffee or glass of water or reality check from my family, i.e. 'Stop writing and get a life!' This is one of the lowest states of awareness in relation to my environment. I do not need to be in a high state of readiness for what may occur in my environment. Similarly, if I were at home reading a newspaper or snoozing in bed, my state of readiness in my environment would be very low, and so it should be: it is called relaxing. However, there are vital higher levels and states of readiness on entering other environments.

Awareness of states or levels of awareness can significantly help your reactions when faced with unexpected incidents.

- **State/Level 1**: White Zone – comfort zone, a comfortable situation, e.g. at home.
- **State/Level 2**: Yellow Zone – everyday situations outside the home, shopping, exercising, coming into contact with other people.
- **State/Level 3**: Orange Zone – more specific tasks, outside the home it may be driving, arriving at work; it is an increased awareness of potential dangers, challenges and conflicts.
- **State/Level 4**: Green Zone – highest state of awareness, being constantly aware of surroundings, e.g. with a client, in a classroom, a police officer on duty, working at home with potentially dangerous equipment.
- **State/Level 5**: Purple Zone – irrational behaviour – instinctual as opposed to rational behaviour takes over → panic → a dangerous state → loss of control.

Tips to help with states or levels

Although the purple zone can be dangerous, the white zone can be just as dangerous. For example, leaving the house still in the White Zone (comfortable, complacent) is probably a major contributory factor to most traffic accidents occurring within one mile of an individual's address. On entering environments that are potentially dangerous in a relaxed and over-confident mode, individuals make themselves highly vulnerable to danger.

Sportsmen and women often experience a similar experience in competition: entering the ring for competition and not having mentally prepared for higher levels of readiness (despite otherwise impressive physical preparations) the athlete can 'freeze', under-perform or (in the case of boxers) suffer a quick knock-out.

Professionals working with the public can encounter 'freeze' behaviours that result in an emotional, instinctive response that is unhelpful for all around them. For example, a psychologist (or teacher or nurse) arrives at work, following an enjoyable lunch break, or perhaps the morning after an indulgent night before. The warm glow of sociability and the cotton-wool feelings of comfort aid the relaxed approach into work (nowadays, my equivalent is heartburn and doziness). On entering work in the White Zone – having luckily survived the journey there! – an incident occurs that requires an immediate response. What may otherwise have been a more measured rational response on a day of greater preparation now becomes a very quick trip through levels 1 to 5. In other words, from comfort to irrational, instinctive – and possibly critical response. From rose-tinted spectacles, to raised voice and occasionally physical confrontation. I think we have all either seen it or felt it.

As a professional, it is essential to stay within the Green Zone to remain rational. To progress and reach the Green Zone it is wise to prepare for each zone: on leaving the house (Yellow Zone) say to yourself 'be aware' and remember to be alert to the drive ahead (Orange Zone). Engage in positive signals to other drivers – allow them into your queue because it helps with engagement of the task. On arrival, acknowledge the people you see, because it helps to prepare you for seeing and speaking to those you may need to communicate with on more complex, challenging levels (transferring from the Orange Zone to the Green Zone).

To enter the Green Zone, look around your environment and address items that help your entry, e.g. arrange chairs, tables – make a room or environment more welcoming for others. Smile and greet others as they enter your Green Zone, enabling you to maintain checking and awareness procedures on those entering.

Police officers are trained to never stay in the White Zone or 'tip over the edge' into the Purple Zone. It is a principle all professionals should aspire to. If we do tip into purple, it can 'rub off' onto the person you are dealing with, subsequently tipping them over into the irrational zone with you. Now it becomes similar to two uncontrollable teenagers facing each other!

In situations where the person arguing has entered purple, an effective strategy is often to (initially) agree or go along with them, and empathise with aggressors. This is usually enough to calm a situation – but if things deteriorate, other approaches may need to come into play.

Defusing danger

First, you need to ensure that you are in control of yourself before you will be able to take control of the situation. The conventional advice is to 'keep calm', but how do you do that? Possibly the most effective way is to take complete control of your breathing. The effect of a greater adrenaline flow is likely to be a faster and uneven breathing. To take control of your breathing, by forcing a more measured, deeper and slower breathing, can impede the wider adrenaline effects and give a greater sense of calmness and control, enabling you to think more clearly and say and do what is more likely to be helpful and appropriate.

Do let people finish what they are saying, even if you have heard it all a dozen times before and you are bored stiff or even if you are sure they are 'trying it on'. Angry and annoyed people want attention, and good listening is one of the best ways of giving someone the attention they demand. Let them know by the way you listen that you are paying full attention to them, you do understand how important it is to them but, while you do sympathise with their plight, you cannot help them, certainly not in the way they may wish, i.e. give them attention, understanding, sympathy, but remain firm.

Don't make promises you cannot keep and that may make things worse for you and/or your colleagues later, e.g. 'Come back at the end of the day and we will see if we can fit you in' – when you know there is no possibility of being able to fit in the person then.

Don't leave an angry or upset individual or relative waiting unnecessarily, perhaps with the intention of 'cooling down', particularly in a public waiting room. People can wind each other up, and people in the waiting room or reception area may encourage the upset person to more serious action – 'Go on, you show them!', 'They should never be allowed to get away with it!', 'You're dead right, they just don't care; they forget it's us that pay their wages!', etc. Suddenly the situation can escalate; the patient or relative cannot climb down easily, compromise or withdraw because a more serious loss of face may be involved now.

Try, if possible, to remove an angry or upset person from an audience; of course, first you will have assessed the possible risk of being alone with that person. Sometimes, in extreme circumstances, it is safer and easier to remove the audience. Be wary of touching the individual when encouraging them to move; many of us, without thinking, will give a light guiding touch to the elbow when asking someone to move or to come with us; this can be perceived as provocative by an angry person and they could possibly react violently.

All common sense, but....

Often there is a resistance by some staff to these sorts of responses because 'they are only common sense!' If only everybody used common sense all the time . . . !' Most staff, experienced and successful at working with the public, will be doing all of these already, so what's new, they may well ask? What we will probably rarely think about is the pattern of our responses to angry people. It is suggested that greater success is likely if we organise our responses according to the following order.

1. **Calm the situation first**

   This includes our body language and calming use of the hands. Listening and encouraging the individual to talk in their own time is important. We want to deal with it in private, away from an audience, and we should be prepared at this stage for anything – even physical violence.

2. **Then show that you understand**

   Maintaining helpful eye-contact and making appropriately encouraging noises show that we care, that we understand and that we want to be as helpful as we can. It might be appropriate to 'personalise' yourself by using your name and making reference to where and who you are.

3. **And now explore the options**

   It is only now that we can really get down to the constructive response to the problem or issue. Now we can clarify precisely what the issue is and explore the alternatives available. It is important that we don't push the individual to the position of total loss of face; we need to explore alternatives, compromises, possible referral, etc.

People's capabilities: responding to highly irrational behaviour

Ask yourself questions about an individual who may become irrational.

- Do they possess the ability to harm me?
- What is their age, fitness, are they 'under the influence'?
- Physical size? – not always a reliable predictor (small build can be misleading)

- Gender?
- Skill, co-ordination?
- Evidence of injuries, e.g. scars, broken nose – also, facial or neck tattoos?
- Disabilities?

Answer to the first question – anybody! Secondly, it is often possible to dominate the conversation with someone younger. Thirdly, differences in size, fitness, gender, skill are not as important as a consideration of what state or zone the other person is in. Note that if an individual stays quiet and withdrawn, it can mean that they are confused or in a state of panic in the Purple Zone, which could well result in unpredictable and dangerous behaviour. Fourthly, injuries, scars or facial tattoos may indicate a rapid resort to violent behaviour – but equally such people may be enthusiastic rugby players or fans of body art! Fifthly, the responses of some individuals who have special needs may become unpredictable if they enter the Purple Zone.

Danger signs to assess violence potential

- Warning signs:
  - direct prolonged eye contact (your eye contact needs to be occasionally broken).
- When angry and highly charged and inside the Purple Zone, the following signs are apparent in the *potential* attacker:
  - facial colour darkens/reddens
  - head held back
  - person maximises their height

- may kick the ground
- large/exaggerated movements towards you, especially the hands
- breathing rate increases
- individual may suddenly stop, then pace the room
- gives indication that *you* are the problem (if you feel you are the cause of the problem, then you should let a colleague take over, ideally a senior one). If an individual maximises their height, then becomes a smaller target with clenched fists – it is a clear warning sign.

Identification of imminent extreme danger signs

- fists clenching and unclenching
- facial colour suddenly pales
- lips tighten over teeth
- head drops forward to protect throat
- eyebrows drop to protect eyes
- hands held *above* the waist
- shoulders tense
- stance changes from square to sideways on
- aggressor breaks their stare and glances at intended target
- if out of reach, the potential attacker may lower their entire body before moving forward to attack.

What the law says about protecting and defending yourself

Those cases that are determined by civil law state:

'If you have an honestly held belief that you or another are in imminent danger then you may use such force that is reasonable and necessary to avert that danger.'

Also:

'Persons may use such force as is reasonable in the prevention of crime.'

[Statute law, Section 3 – Criminal Law Act 1967]

Remember – at all times you must act within the law – you are responsible for your own actions.

Finally, breakaway techniques –

Arms | | rather than ═

It is best to hold arms vertically when protecting yourself, simultaneously twisting your trunk sideways to deflect blows.

However, it can be helpful to remember the mnemonic LEAPS in an attempt to avoid confrontations/conflict:

Listen

Emphasise

Ask

Paraphrase

Summarise

So far, we have looked at threatening aspects of staying safe. To be aware and act on your awareness reduces the likelihood of being physically, or emotionally, harmed.

What follows is a preventative approach to staying safe in work. In other words, an approach to avoid what has been described so far.

Section 2
Conflict resolution: the preventative approach

Diagram 2.1: Conflict resolution: importance of image and body language

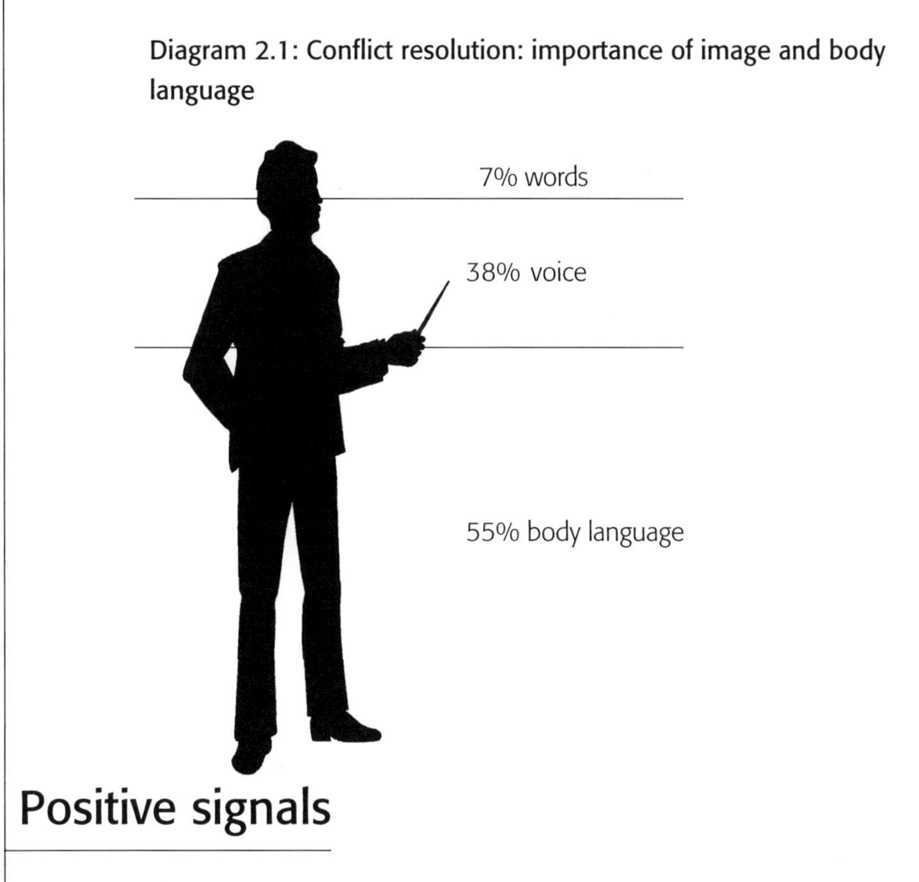

Positive signals

The usefulness of understanding body language is not only in terms of reading other people's but also to help us reassess our own, and the impact this could have on potentially difficult situations.

It must be stressed that presenting a positive image is not suggesting that this will avert every potentially difficult situation. Where a mixed message is sent, the non-verbal message will be the strongest so it is important that our non-verbal behaviour complements what we say.

Body language can be broken down into different components.

Eye contact

When we look at someone we are inviting them to interact with us. As we grow up we pick up the rules of eye contact, from cultural influences, models of family, teachers, etc.

As a general rule confident eye contact is direct but the contact must occasionally be moved away to give the other person some relief. Too much eye contact quickly becomes an aggressive stare that is uncomfortable for the receiver.

Withdrawing eye contact by lowering the eyes is often interpreted as a signal of submission, although this often has to be interpreted in the context of cultural differences.

Hand Gestures

Possibly one of the strongest non-verbal gestures is the hand gesture. There are three main positions.

- An open upturned palm is used as signalling you are non-threatening and carrying no weapons – using open-palm gestures may help people calm down.
- When the palm is downwards it is more likely to be interpreted as an order that you are indicating you have more power in the situation.

- The pointing finger is often widely recognised as an aggressive use of a weapon. The finger-tapping gesture often indicates an impatience with what is being said, and often precedes an escalation of aggression.

Arm barriers

Crossed arms are often interpreted as a negative response, it may well be that the arms are crossed because the person is cold, but nonetheless it may make people feel you are negative towards them or what you are saying.

When a cluster of gestures exist, maybe crossed arms and clenched fists, this may indicate a hostile and aggressive attitude.

Stance

A confident stance should have both feet firmly grounded and centred, partly because of the non-verbal signals of being in control, but also for personal safety. If someone is standing with ankles crossed or weight on one side, they would, in an aggressive situation, be very easy to overbalance.

Personal space

Everyone has a space around them which they feel should not be entered by other people unless invited. This space varies from individual to individual, it has been estimated that this space is between 12 and 18 inches for most people (the intimacy space). If someone enters that space uninvited we feel threatened. If someone is showing signs of aggression, moving into their personal space is likely to make the situation worse.

If someone has invaded your space and you are beginning to feel uncomfortable it is important to try to redress the situation while looking as confident as possible.

A positive step backward, keeping both feet centred, is often seen as a clear signal that the other person has come too close, especially if accompanied by direct eye contact. If you cannot move backwards a clear step sideways is as effective.

If your space is invaded while you are sitting this will be even more threatening and, again depending on the situation, you could ask the person to move further away or stand up and move out of the situation.

Use of voice

- Avoid words such as 'can't', 'should', 'must', 'ought to', 'have to'.
- Keep voice neutral, calm.
- Ask how? instead of why?

Anger

The sorts of behaviour that can trigger aggressive or violent reactions could include, for example:

- talking down to people
- patronising them
- telling people they are wrong to feel the way they do
- standing on your official dignity
- trivialising people's concern, upset, frustration, problem
- using the wrong form of address or the wrong name

- expressing assumptions
- ridiculing
- using organisational jargon.

One method I have used for ten years to keep clients, myself and colleagues safe is called the **Support and Risk Assessment (SaRA)**. I am grateful to Geoff Shepherd for transcribing the approach, which was subsequently published in 2001 *(Shepherd and The Independent Psychological Service, 2001).*

The SaRA form is provided in the Appendix; a complete assessment guide is available – see References at the end of this book. The main essence of SaRA is described below, and has been used with vulnerable and challenging clients from schools, colleges, prisons, secure units, hospitals, voluntary agencies, youth offending teams and care provisions.

Support and Risk Assessment (SaRA)

Aims

To assess and manage student- or client-related risk to provide a safe, successful environment for students, clients, staff, public and the service.

What are the key components of the programme?

- defining the risks
- a framework for assessment

- creating a supportive environment
- risk management.

What risks are being assessed?

- Things that could put the student or client at risk.
- Things that could put other students or clients, staff or the public at risk.
- Things that put the service and its property at risk.

Some examples:

- *lack of confidence* can be very significant for students who have mental health needs. One prospective adult student at a college said he was worried that tutors might make him 'feel even more inferior than I am!'
- many people take *medicines* but their effectiveness can vary and can affect mood or behaviour
- some people have *difficulty with change.* Changes in room layout or patterns of activities may provoke reactions from some people.

How is the SaRA used?

- *A risk rating score* will indicate the specific areas and level of risk:
  1 – 3: low level
  4 – 6: medium level
  7 – 9: high level.
- The risk assessment will guide students or clients to *appropriate choices* of course or activity.

- Extensive *information* is collected to support the student or client and staff.
- *Support guidelines* will be agreed and made available to tutors or staff working with the student or client.
- *The SaRA programme includes students or clients* in the ongoing assessment of risk and agreeing guidelines for support.
- Tutors or staff will have *key support information* and a *named person* to consult for advice, guidance and future information.
- Provides a powerful risk management tool.
- Provides specific information to inform the development of support resources and programmes.
- Provides a focus for multi-agency working in relation to each student.

Example SaRA forms (the information base form and the rating scale form) are provided in the Appendix (page 43).

Principles

Risk assessment:

- is a positive tool to inform and guide students or clients
- engages the student or client in a realistic dialogue about their strengths and weaknesses
- requires each agency to offer a range of support in response to identified needs
- should be a pro-active tool preventing unnecessary problems arising
- needs to be 'owned' by the student or client, resulting in an effective and mutually positive partnership aimed at success

- is confidential to the assessor and the student or client with information going to others only as necessary.

Risk rating scores and support responses can be available to other key people following discussion and agreement of the assessor and the student or client.

Key approaches

The assessor should:

- make sure that the student or client is involved and informed
- involve other people who can contribute relevant information
- be straightforward and honest, keeping the process as a positive dialogue
- encourage the student or client to offer their own views of the risks and possible responses
- present the process as useful to students or clients as a way to identify and get the support they need to be successful
- be clear that sometimes it will not be possible to do what the student or client wants
- have a close partnership with the support resources of the service
- make sure that the student or client is involved and informed
- ensure that there is an effective system for keeping assessments in a confidential place
- ensure that all relevant staff have the opportunity to know which students or clients have completed SaRA
- be clear that unwillingness to engage with the SaRA process means that the student or client may prohibit themselves from access to a course of study, activity or training opportunity.

The client–partner approach

The first meeting of the assessor and the student or client is very important. The message that the assessor needs to get across is: *'The school/college/ scheme want you to be part of it. Our job is to look at your goals and explore what will get in the way and what will help you succeed. Then we will see how we can support you to make sure that the things that might go wrong, don't go wrong.'*

The dialogue of the assessment is primarily between the assessor and the student or client. The important contributions from key others are amplifications, checks and guidance.

The risk rating score should record the student or client's own view of their rating as well as the assessor's agreed final score.

Information search

Engaging the student or client and key professionals in a realistic dialogue about their strengths and weaknesses

- The information-gathering stage should use key professionals identified in the network of involvement.
- Students who have a number of professionals in their network of involvement should have one of those professionals accompany them to the assessment interview.
- The student or client should be asked for permission to request information and to involve key professionals in the assessment interview and process.
- The information base form can be sent out to gather information from key professionals who are not attending assessment meetings.

- If the student/client does not give permission for contact to be made the assessor should explore the reasons with them. The assessor will need to make a judgement about the potential significance of missing information and increase the risk rating accordingly.

- The assessment interview explores the student's or client's social and educational background using the information base forms. Particular attention should be paid to the personal perspective, views and feelings of the subject about their history and future.

- A central piece of information is *'What does the student/client really want to achieve?'*

- A key insight is the subject's view of *'What might happen that will make me fail?'*

The risk assessment process

Using the SaRA baseline form* and the rating scale with students or clients

The first phase of any meeting or discussion with a potential student or client aims to establish trust and confidence between the interviewer and the subject.

Talk initially without an obvious, formal schedule about things that are not stressful for the subject and in as stress free an environment as possible:

'What would you like to drink...tea...coffee?'

'Where do you live?'

'What are you doing at the moment?'

* See example forms in Appendix (page 43)

'What are you hoping to do in the near future? ...and later on? (e.g. work?)

Introducing the risk assessment as *'how we find out what might get in the way of you being successful and what support might be needed to make sure that you are successful'* establishes two fundamental ideas:

a. It is in the student's or client's interest to discuss all those things that may become difficulties in the future.

b. Risks and needs are met and matched by a variety of support.

1. Put questions in a clear and non-threatening way.

   For example, a question *'So what things make you feel good, positive or confident?'** will provide important insight into how support can be designed to produce positive outcomes.

   Similarly, a question *'So what makes you really angry or upset?'** might lead to information about how anger is managed, what the negative triggers are, whether medication is being used to control mood, or whether the person has difficulty in accepting direction or perceived criticism.

2. Explain the rating system and how it links to levels of support. Ask the subject for their view of the right SARA score for them. Indicate your own preliminary score and explain any differences. (An alternative approach to arriving at a SARA score is that all those attending the SARA meeting give a rating score at the end – including the client – and the average is taken as the SARA score.)

3. The student or client should be told that your report will be completed as soon as possible and that they will be given a copy at your next meeting so they can add to or amend it.

4. Discuss and agree the support response for each area of concern. The support response will need to be discussed with the relevant service manager so that the support response package can be confirmed or amended with the subject at the next meeting.

* It is important to note all that is said in response to these questions

5. The frequency of meetings to monitor progress needs to be set and effectively communicated to the student or client.

6. Apart from reviewing, the assessor does not need to be the focal person after this point but should remain until the responsibility for implementing the support package is accepted by a specific other person in the network of involvement.

7. The assessor should discuss and agree with the subject who else needs to have copies of the assessment and who needs copies of the support response. They should agree who will give or send this information.

8. Copies of assessments and support response plans should be held in the student's personal file, the assessor's file and a central support service file. All should be protected by confidentiality protocols.

9. The student or client should be assured that the information held is only to be used for advice and guidance to the subject and in drawing up support response packages.

Programme negotiation

Making the deal

The 'deal' is an agreement to enter into a plan of activity with a student or client who requires support to avoid possible risk. It requires a number of things to be present:

1. The real goals and ambitions of the client.

2. An agreement by the student or client to engage in the SaRA process.

3. An identification of the interests and strengths of the client.

4. Agreed areas of risk and support.

5. An agreed strategy for progress, step by step and with a timeframe.

6. Clear boundaries for the student or client about possible unacceptable performance or behaviour.

7. One person responsible for maintaining regular contact and an overview of the progress of the student or client.

8. A focus on succeeding through a supportive partnership.

9. Celebration of success and small steps towards success.

10. A commitment by staff to work with the positive ambitions and qualities of the student or client.

11. A commitment to regular meeting in semi-informal settings to discuss progress and reinforce 'the deal'.

Section 3: Risk

Definitions of risk

'the probability of an event'

'...embraces categories of suicide and self-harm and of severe self-neglect' (NHS Executive 1994)

'...some definitions have sought to emphasise a balance by introducing the potential for beneficial as well as harmful outcomes' (Carlson 1995)

'...a further addition of 'stated timescale' extends to a definition of the potential predictive value of likely outcomes'

A working definition may be:

'Risk is the likelihood of an identified behaviour occurring in response to changing personal circumstances. The outcomes are more frequently harmful for self or others, though occasionally may have a beneficial aim in pursuit of a positive challenge.'

Risk behaviours

- aggression
- violence

- suicide and self-harm
- severe self-neglect
- others.

Who is at risk?

You are more at risk from crime if you are:

- **a young man**: this group is at the greatest risk of robbery and assault
- **a man**: men are more at risk from crime overall than women
- **Afro-Caribbean**: this group is more likely to be victims of crime than white people
- **Asian**: Asian people are more likely to suffer vandalism, personal theft and victimisation.

Definitions of violence

Physical violence is:

- assault causing death
- assault causing serious physical injury
- assault causing minor injuries
- kicking
- biting

- punching
- use of weapons
- use of missiles
- spitting
- scratching
- sexual assault.

Non-physical violence is:

- verbal abuse
- racial or sexual abuse
- threats – with or without weapons
- physical posturing
- threatening gestures
- abusive phone calls
- threatening use of dogs
- harassment in all forms
- swearing
- shouting
- name calling
- bullying
- insults

- innuendo
- deliberate silence.

oiding aggression and violence

- use open discussion
- allow plenty of time
- use open questions in a non-confrontational and non-judgemental way
- move to a problem solving-solving approach
- maintain a calm manner but be cautious about initiating touch too soon as it may be seen as invading personal space or cultural norms
- maintain a calm pace and tone to your voice in the face of verbal aggression
- assess threat for immediacy and don't be overwhelmed... it is not likely to happen immediately
- if you think threats are immediate and likely to be acted upon... leave and seek help... do not retaliate with counter threats
- control and restraint techniques should only be used if staff have had the appropriate and updated training... this is the last resort.

And finally...

If you are ever verbally or physically abused, it is essential to speak with a trusted colleague (or other trusted individuals) as soon after the incident as possible. Research into Post-Traumatic Stress Disorder (PTSD) has demonstrated that the more quickly someone can release verbally what has happened to them, the less chance there is of experiencing a post-incident effect, e.g. continually reliving the experience, sleeplessness, panic, etc. When 'releasing' the events to another, try to provide the facts, the feelings, what you saw, what you heard, any tastes, odours or smells in the area it occurred, any external sounds, any textures or touch sensations. In other words, by later connecting the events with all your senses, you are less likely to experience an unpleasant reliving of the experience. It is then you can plan what to do to address what has happened and prevent it occurring again.

References

Piven, J. and Borgenicht, D., (2001), *The Worst-Case Scenario Survival Handbook: Travel*, San Francisco: Chronicle Books.

Shepherd, G. and The Independent Psychological Service (2001), *SaRA: Support and Risk Assessment*, available from Geoff Shepherd, Community Living Options, 34 Seymour Road, Bath BA1 6DY.
(Price £27.50; unlimited photocopying for each original organisational purchase.)

Appendix

Support and Risk Assessment (SaRA) forms

Information base form

Student/client name: ______________________________

Address: ______________________________

____

Date of birth: ______________________________

Course/activity: ______________________________

From/to: ______________________________

Completed by: ______________________________

The following people have contributed to this information base:

1. ______________________ Role: ______________ Tel: ____________
2. ______________________ Role: ______________ Tel: ____________
3. ______________________ Role: ______________ Tel: ____________
4. ______________________ Role: ______________ Tel: ____________
5. ______________________ Role: ______________ Tel: ____________
6. ______________________ Role: ______________ Tel: ____________
7. ______________________ Role: ______________ Tel: ____________
8. ______________________ Role: ______________ Tel: ____________

What reports or other information do you have access to?

Do you need other information to form a complete picture?

Who has contributed to the information in this baseline information report?

What is the background of the person?

(Include family, education, post-school activity, key achievements and difficulties)

What work has this person done?

What goals or ambitions do they have for the future?

What are their positive triggers?

What are their negative triggers?

Areas of concern needing a support response

1. ____
____
2. ____
____
3. ____
____
4. ____
____
5. ____
____
6. ____
____
7. ____
____
8. ____
____
9. ____
____
10. ____
____

Are any other specialist reports or assessments needed?

☐ Dyslexia ☐ Speech therapy ☐ Psychology

Other:

Who will draw up the support response?

1. ____________________ Tel: ____________________
2. ____________________ Tel: ____________________
3. ____________________ Tel: ____________________

Describe the specific support response the person will need to be successful and safe

1. ______________________________
____
2. ______________________________
____
3. ______________________________
____
4. ______________________________
____
5. ______________________________
____
6. ______________________________
____
7. ______________________________
____
8. ______________________________
____
9. ______________________________
____
10. ______________________________
____

Complete the rating scale form to determine the SaRA score

1 – 3 low:	occasional support
4 – 6 medium:	regular or significant one-to-one support
7 – 9 high:	needing one-to-one support all the time

SaRA score

Date for assessment review: __________

Completed by: __________

Date: __________

Rating scale form

Using the rating scale form

- Use the information collected in the 'information base' form to complete the 'rating scale.'
- The rating scale summarises the level of risk for each person assessed and acts as a standardising framework.
- It is recommended that the completed assessment is countersigned by one of the external agencies.
- Use the blank spaces on the 'description of risk factor' page to enter factors particular to your student or client and your setting and environment.
- The completed rating scale must be reviewed and updated on an agreed time schedule. Initial assessments are best reviewed within three months.

SaRA rating scale form

Student/client name: ______________________

Address: ______________________

____

Date of birth: ______________________

Next review date:

Course/activity: ______________________

From/to: ______________________

Other key reports:

-
-
-

Support information to:

-
-
-

Overall risk rating score

Assessor: ______________________

Date: ______________________

Contact person: ______________________

Tel: ______________________

External agency:

Date:

Name:

Signed:

Description of risk factor

Factor range 7–9	Yes	No	Notes
•			
•			
•			
• Evidence of injuring others (last six months)			
• Incidence of damage to property (during previous year)			
• Large changes in behaviour or mood			
• Evidence of unsettled medication programme			

Factor range 4–6	Yes	No	Notes
•			
•			
• Requires highly-structured programme			
• Medication to control mood or behaviour			
• Difficulty in maintaining regular attendance			
• Difficulty in making appropriate relationships			
• Fear of noise or crowds causing anxiety			

Factor range 1–3	Yes	No	Notes
•			
• Inability to cope with changes to routine			
• Finds criticism or direction difficult to accept			
• Difficulty in retaining key information			
• Low ability to deal with stress			
• Low confidence and self-esteem			
• Limited attention span in learning situations